Data Quality in the Age of AI

Building a foundation for AI strategy and data culture

Andrew Jones

‹packt›

Data Quality in the Age of AI

Publisher: Vishal Bodwani
Product Manager: Sathya Mohan
Lead Development Editors: Siddhant Jain and Oorja Mishra
Development Editor: Afzal Shaikh
Copy Editor: Safis Editing
Proofreader: Safis Editing
Project Coordinator: Yash Basil
Production Designer: Deepak Chavan
First published: May 2024
Production reference: 1300724

Published by Packt Publishing Ltd.
Grosvenor House, 11 St Paul's Square, Birmingham, B3 1RB
ISBN 978-1-80512-143-5

www.packt.com

Executive summary

Organizations worldwide are eager to capitalize on recent advancements in AI and harness its newly available capabilities. However, even the most sophisticated models require high-quality data to be effective. This report provides practical and actionable steps to enhance your data quality, a crucial factor in the age of AI.

This definitive resource delves into the following key areas:

- **Understanding data quality**: Learn to define, assess, and measure your data quality, providing a clear view of your current state and areas for improvement.

- **Improving data quality at source**: Gain practical advice for enhancing data quality during its creation, where it is most cost-effective and impactful.

- **Case studies**: Explore real-world examples that demonstrate the significant positive impact of focusing on data quality within organizations.

- **A culture that values quality**: Foster a data culture that prioritizes quality, treats data as a product, and embeds data governance throughout the data lifecycle.

Data Quality in the Age of AI offers crucial insights and actionable advice for enhancing and maintaining data quality at the point of production, where it is most cost-effective and impactful, embedding these improvements into your organization's data culture. This ensures your organization is well-prepared to leverage the latest AI capabilities swiftly and effectively.

Target audience

This report is aimed at data leaders, decision makers, and other key stakeholders responsible for developing data strategies that directly support their organization's AI objectives.

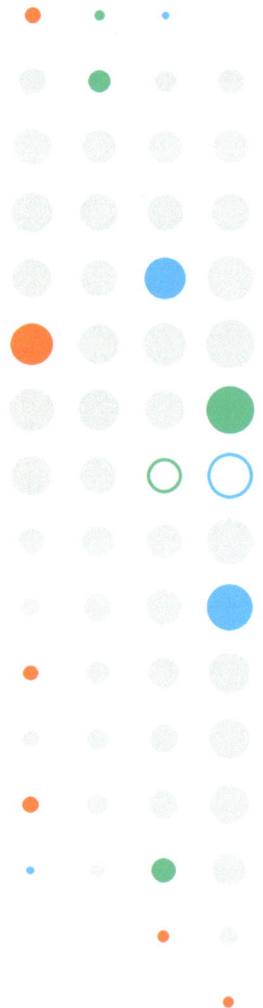

Contents

Understanding data quality

Organizations everywhere are looking to refresh their data strategies to take advantage of the recent advances in AI and position themselves to capitalize on the newly available capabilities. However, no matter how advanced AI models get, they'll rely on *quality data* to be effective. New research from Informatica concluded that 42% of data leaders cite data quality as their top obstacle, while 40% highlight data privacy and governance, and 38% point to AI ethics as significant challenges.[1]

The quality of your data is not going to improve without a strategy that clearly articulates why doing so is *critical* for your organization to achieve its AI goals. This report gives you everything you need to clearly articulate why you need to focus on the quality of your data as a key part of your overall data strategy. You will also discover some direct and applicable actions you can take to make data quality the foundation of your data culture.

After a clear introduction to data quality and what determines it, you will learn how to measure the quality of data in your organization, enabling you to understand the scale of your problem and its impact. This report will also discuss how you can improve the quality of your data at the source and introduce two emerging architecture patterns

"The quality of your data is not going to improve without a strategy that clearly articulates why doing so is *critical* for your organization to achieve its AI goals."

to support this change. Finally, you'll learn to embed data quality into your culture, ensuring your organization will benefit from quality data well into the future.

Implementing the recommended actions in this report can help position your organization to fully capitalize on advances in AI, significantly enhancing the value you can derive from your data.

Before you consider the solutions, let's discuss what is *data quality*. What exactly constitutes data quality, and what factors determine it. Moreover, why is it crucial, particularly in support of an organization's AI strategy, and what are the repercussions of neglecting it.

Defining data quality

The question of quality data, and the consequences of using it, is a topic as old as data itself. And yet, as old a problem as it is, data quality is often hard to define. There are many definitions being used, often interchangeably. Some emphasize different dimensions of the data itself, like its *completeness*, *accuracy*, and *timeliness*. But these dimensions are not sufficient to tell whether the data is of good quality.

Other definitions consider whether data is fit for *purpose*— which is rather broad—but it's a better definition and the one that's naturally used every time one looks at a dataset. When determining whether data is fit for purpose, one should also ask whether they *trust* the data enough to make a decision, take an action, or build on top of it.

"

When determining whether data is fit for purpose, one should also ask whether they *trust* the data enough to make a decision, take an action, or build on top of it."

Organizations today face significant challenges in extracting business value due to poor data quality. By being deliberate about how the business creates, manages, and provides data to other parts of the organization, and being clear on the responsibilities data producers and consumers have, organziations can transform how they use quality data to meet their strategic goals—including the deployment of AI.

Assessing data quality

Determining whether your organization's data is trustworthy and fit for its intended purpose hinges on aligning dataset *expectations* with actual needs. Often, these expectations are not set by those generating the data, and in their absence, users tend to make optimistic assumptions. However, when reality falls short of these expectations, trust in both the dataset and the data erodes. Rebuilding trust becomes exceedingly challenging once it's lost.

Organizations can set expectations by assigning and monitoring different dimensions of data. The figure here shows some useful dimensions for capturing the expectations of a dataset.

The importance of these dimensions varies depending on the dataset. It's important to note that organizations shouldn't expect perfection in each dimension. In fact, the required level varies greatly based on the intended use. For instance, suppose you're examining data from a

Dimensions of data quality

Usefulness
Intrinsic value of data

Accuracy
Reflectiveness of reality

Validity
How data meets criteria

Uniqueness
Non-duplicative rows and tables

two-week experiment on your website to assess if it improved the conversion rate. In this scenario, while striving for completeness and accuracy, it's acceptable if the data isn't 100% perfect. However, you wouldn't want it to fall below a certain threshold—perhaps around 80% completeness and accuracy would suffice for this analysis.

In terms of timeliness, having data available up to the end of the previous day might be acceptable. However, if the data is a week old, you may be missing half the data, which could significantly impact the results of your analysis. However, it's important to acknowledge that there may be another user of this data who may have different requirements. For instance, customers involved in the experiment who need direct contact from the support team might necessitate a dataset that is 100% complete up to the experiment's conclusion, with accurate email addresses. Once more, you need to ensure that the data is *fit for its intended purpose*.

Each dimension is initially set when the data is first generated. If, at the time of generation, you're not capturing the records accurately or you're missing records, there's nothing you can do downstream to add that data back in. For example, if the data is generated daily, nothing you do downstream can make it timelier.

Therefore, *data quality must be improved at the source*. Any attempt to improve its quality anywhere else severely limits what it can achieve.

Completeness
Robustness of data

Consistency
Uniformity in values and structure

Timeliness
Readiness of data

Unlocking AI's potential with data

Recent advances in AI and increased accessibility to machine learning models that power them have got many organizations excited about how they can be applied to create more value. In fact, in a recent survey by Dataiku and Databricks, 64% of respondents said that they were "likely" or "very likely" to use generative AI for their business over the next year,[2] while a report from Segment found that 92% companies are using AI-driven personalization to drive business growth.[3]

However, *AI is only as good as the data behind it.* No amount of tuning a model will help if the data is of poor quality, which is why the same survey from Dataiku and Databricks identified *lack of quality data* as the primary obstacle to generating value. An inferior model with superior data will always outperform a superior model with inferior data.

"

An inferior model with superior data will always outperform a superior model with inferior data."

9/10

companies use AI-driven personalization for growth

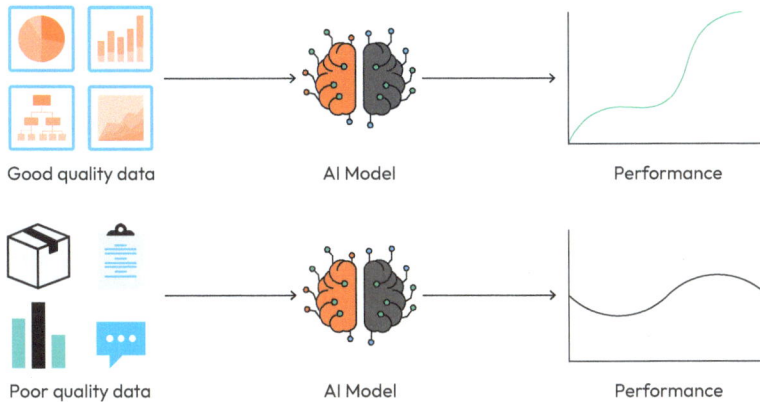

Impact of data quality on AI model performance

There are other data issues that can affect the quality of the model, such as data drift, but they are not directly *quality* issues—they're part of what's required to build a model that still performs well as data patterns change over time.

Not only is the outcome determined by the quality of the data, but so too is the effort required to build and deploy a model. A 2022 survey from Anaconda found that data professionals spend approximately 38% of their time preparing and cleansing data.[4] That is time wasted due to the poor quality of data that is available to them.

In 2022, Unity Software reported that some bad data ingested from one of their large customers affected a machine learning model responsible for placing ads and resulted in approximately $100 million in costs to the business[5] due to the lost ad revenue and the work required to recover from the incident. This example illustrates just one aspect of the costs associated with poor data quality, but there are numerous others to consider. Let's explore some other examples next.

High cost of poor data quality

There have been many surveys and reports that try to quantify the cost of poor data quality to an organization. Quintly estimates poor data costs an average loss of $15 million per year, while IBM estimated it cost US businesses $3.1 trillion in 2016 alone.[6] Meanwhile, in a recent survey commissioned by Monte Carlo, more than half of the respondents reported that poor data quality is impacting 25% or more of their revenue, while the average from all respondents was 31%.[7]

These findings outline both the *direct* and *indirect* ramifications of poor data quality. *Direct costs* are those that have an impact in the short term and include the breakdown of an operational process, the downtime of a data-driven product, and the loss of productivity for employees. *Indirect costs* are those that have an impact over the long term, impacting decision making at all levels of an organization, ultimately harming business performance.

Poor data quality can also lead to increased risk for businesses. Failing to adhere to regulations regarding the storage and utilization of data can result in potential fines and negative media exposure. Even seemingly minor infractions, such as sending marketing emails to individuals who have opted out, can reduce trust in the brand and expose it to regulatory fines.

$15 million

Average losses incurred due to poor data quality

25%

Or more revenue impacted by poor data quality

Measuring the quality of your data

Before you start working on projects and putting in place programs to improve data quality, you should first measure the quality of your data. This step is crucial for justifying investments and establishing success metrics.

There are three ways you can measure the quality of your data:

- Gather feedback from data consumers about how much they *trust* the data they are using.

- Run tests and profile, or continuously *observe* the quality of data.

- Start treating issues with the data as an *incident* to reinforce the impact of poor-quality data.

Trust

Most organizations believe their data is unreliable. 82% say data quality concerns are a barrier to data integration projects,[8] whereas around 80% of business executives do not trust their data.[9] Trust comes from the quality of the data and whether it is fit for its purpose. Therefore, the level of trust data consumers place in the data they utilize serves as a good measure of an organization's data quality—be it for a specific dataset or indicative of a broader lack of confidence in the organization's data integrity.

One way to measure trust is by conducting regular surveys and interviews with data consumers. Here are some questions you can pose:

- How do you assess the quality of the data? Can you use it directly, or do you often need to validate it?

- Have you faced any data quality issues that affected your trust in this data? Can you describe those issues?

- How often do you use data for your daily tasks? What's preventing you from making use of more data?

- Where have you recently used data successfully to generate business value? What data was it, and how did you use it?

These surveys offer another valuable data point for measuring the quality of your data. They provide you a roster of engaged data consumers whom you can schedule interviews with to gain further insights into how data quality impacts their daily tasks.

With this, you now have an understanding of how people *feel* when using data. Now, let's explore how you can gain better visibility into the specific data quality issues you're facing and where they're occurring.

Observability

Gaining greater visibility into the scope of data quality problems is an important step to take before starting any project or program. This step offers valuable insights into the specific issues at hand, their origins, and their repercussions.

> **"**
> Gaining greater visibility into the scope of data quality problems is an important step to take before starting any project or program."

It's easy to get started by creating some data quality checks and running them as part of your data pipeline using open source tools such as Soda, Great Expectations, and DQOps. These checks can target the most important data sources and include the following:

* Freshness

* Duplications

* Missing values

* Values out of the expected range

It's also possible to profile your data, giving you a one-off report about the data—what values are present, any outliers that exist, and other statistical characteristics. Great Expectations supports data profiling, as do dedicated tools such as Google Cloud Dataplex.

You can also consider onboarding a data observability solution. These solutions continuously monitor each of your data assets and alert you when anomalies are detected in your data. For example, if there was a sudden increase in missing values, or if the data arrived later than expected. Popular vendors in this category include Monte Carlo, Metaplane, and Sifflet.

Whichever approach you choose, these checks can only run after the data has been generated, by which time the impact of the poor-quality data may already have been felt by those depending on that data.

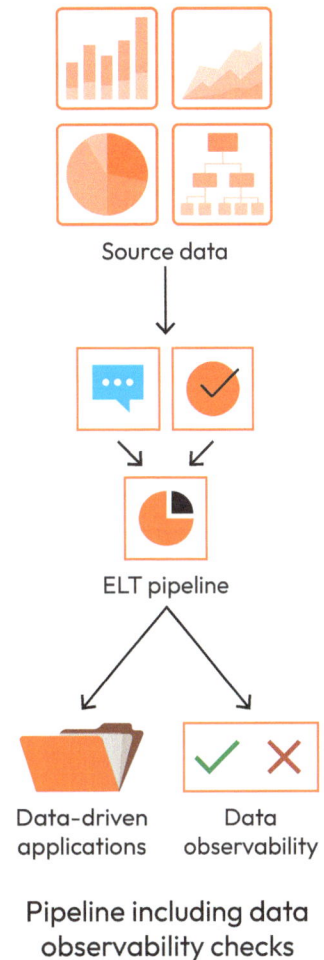

Source data

ELT pipeline

Data-driven applications

Data observability

Pipeline including data observability checks

While these tools won't directly enhance the quality of your data, they provide invaluable visibility into its current state. These insights are valuable for strategizing solutions for solving data issues at source. As you implement these checks and/or deploy an observability solution, you'll start receiving alerts when data issues are detected. Treating these data issues as *incidents* increases the visibility of data quality problems and the impact they have on your business.

Data incidents

When you're working with data at scale and/or speed, it's inevitable that things will go wrong. This is well accepted in terms of software, but the same goes for data too. If you want your organization to understand the importance of poor data quality—and the issues caused by it—then you need to start treating your data incidents in the same way good software engineering teams do. Exactly what this looks like depends on your organization, but it should include the following:

- Raising an incident as soon as an issue is detected and assessing the severity in terms of its potential business impact.

- Working together with relevant teams, including those whose work may have unknowingly caused the incident, to resolve issues as quickly as possible.

- Giving regular updates to stakeholders so they know whether the data they are using is affected by the incident.

66

Treating these data issues as *incidents* increases visibility of data quality problems and the impact they have on your business."

- Keeping track of the investigation and the actions you took during the incident as you work to resolve it.

- Once an incident has been resolved, it's important to analyze the incident through a postmortem process to understand the root cause, capture the key takeaways, and take action to prevent similar failures in future.

Popular incident management tools are PagerDuty, incident.io, and FireHydrant. Each of these tools can guide you through these steps while following best practices. They also help you keep track of incidents and their actions, either within the tool or by integrating with incident management tools such as Jira. Incident management allows organizations to improve the visibility of issues they're facing and the impact they have on the business.

Utilize the postmortem process as an opportunity to engage with data producers and highlight the consequences of data issues. It's crucial, however, to avoid apportioning blame during this review. Instead, focus on fostering relationships and starting constructive dialogues to enhance collaboration and collective improvement efforts.

Furthermore, following an incident process for data incidents provides organizations with qualitative data on the number and nature of the incidents. This information can be used to identify the types of incidents, the affected data, and the applications they impact. Such insights serve to construct a compelling business case for proactively addressing these incidents through investments in data quality enhancements.

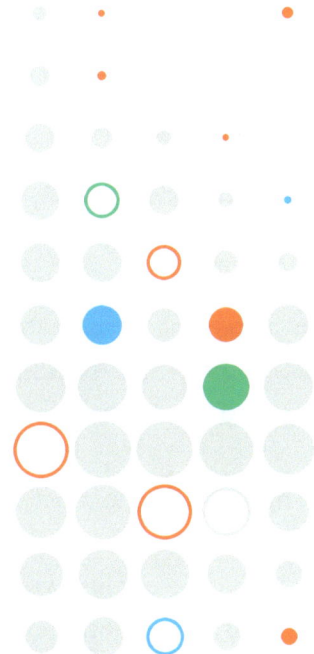

Improving data quality at the source

An Experian report conducted in 2021 found that 95% of business leaders reported a negative impact on their business due to poor quality data.[10] This underscores the necessity for proactive measures to improve the quality of the data.

Data quality can only be improved at *source*. If the data source fails to capture information accurately, rectifying it later becomes futile. Similarly, inaccessible data sources can affect user access. If data is delivered infrequently, its timeliness cannot be retroactively improved. Likewise, if data sets are incomplete at the source, there's nothing you can do to make them complete later.

You can try to work around some of these data quality issues downstream, typically in your data pipelines. For example, you could impute missing values using averages, the most common values, or machine learning algorithms, but these may be inaccurate, introduce bias, and be expensive to compute. This also adds complexity to the data pipelines, decreasing their maintainability and reliability.

The source is the most cost-effective location to identify and address data quality concerns. This is known as the 1:10:100 rule of data quality, which was developed by George Labovitz and Yu Sang Chang back in 1992 and states the following:

95%

business leaders report negative impact to business due to poor data quality

- The cost of preventing poor data quality at the source is $1 per record

- The cost of remediation after it is created is $10 per record

- The cost of failure (in other words, doing nothing) is $100 per record

$1 for prevention

$10 for correction

$100 for doing nothing

The 1:10:100 rule explaining the cost of poor data quality

Preventing poor-quality data from leaving the source can help cut costs significantly. Doing so confines the impact of data issues, sparing multiple systems from costly remediation efforts across various teams. Moreover, it prevents potential user dissatisfaction or negative repercussions on your organization's reputation, which can incur expenses related to user churn and reputational damage.

By embracing the notion that data quality can only be improved at source, you acknowledge that the responsibility for data quality rests with the data producers. Data producers possess the full context of the data and how it is captured or generated, along with the authority to enhance data quality by refining their services.

> **By embracing the notion that data quality can only be improved at source, you acknowledge that the responsibility for data quality rests with the data producers."**

However, data producers are often occupied with existing responsibilities. Therefore, for them to take on additional tasks, you must address two key aspects:

- **Incentivization:** Offering incentives to data producers highlights the significance of their role in preserving data quality and encourages active engagement.
- **Architectural support:** It's essential to furnish data producers with an appropriate architecture to guide them in generating quality data effortlessly.

Let's delve deeper into these strategies.

Incentivizing data producers

For data producers to take on the responsibility of providing better quality data—and to do it well—they need to have incentives that align with their work. There are many ways to do this. Every organization that has grown beyond a start-up needs a way of incentivizing multiple teams to work together to build something of value for the business.

One effective approach is a top-down strategy. By aligning strategic objectives that rely on the creation and utilization of high-quality data, organizations can optimize their structure to facilitate collaboration among relevant teams. Additionally, employing **key performance indicators (KPIs)** and other prioritization methods can ensure that teams are held responsible for their contributions toward achieving this goal.

For example, if you're tracking data incidents, you could create KPIs around the number of incidents, their severity, and how common the root causes are. While you will never get the number of incidents to zero, this is a great place to start if you currently have far too many data incidents.

The other approach is more localized. You can encourage data producers and consumers to collaborate by raising awareness, building relationships, and designing tools and architectures that require this collaboration and provide space for it to grow.

Whichever approach you take, you'll need to support data producers, so they can produce and make available quality data. There are various compelling architecture patterns that have emerged recently that aim to provide this support.

Decentralizing your data: Quality by design

There are two emerging, complementary architecture patterns that aim to support a move to a more decentralized data culture to drive the production, management, and use of quality data that provides value to the business: *data contracts* and *data mesh*.

Data contracts

Data contracts aim to facilitate and encourage the production of quality data through the use of well-defined *interfaces*. These interfaces act as the agreement between the data generator and the data consumer, capturing the expectations that have been set around the data.

A data contract functions similar to an API, serving as a reliable interface for data. While APIs are employed by software engineering teams to offer quality interfaces to their services, data contracts fulfill this role for data. The primary distinction lies in the type of interface provided. APIs typically offer a REST API or similar interface accessible over the internet. In contrast, data interfaces are generally a table within a data warehouse such as BigQuery or Snowflake, or as data streams, such as Kafka or a Google Cloud Pub/Sub topic.

Setting up the tooling for data contracts is a straightforward process and is deployable with minimal effort. A basic solution can be developed and implemented by a small team within a few weeks. All that's necessary is the capability to define the data contract and establish its interface. Subsequently, data producers can input data through the interface, while data consumers can retrieve data from it.

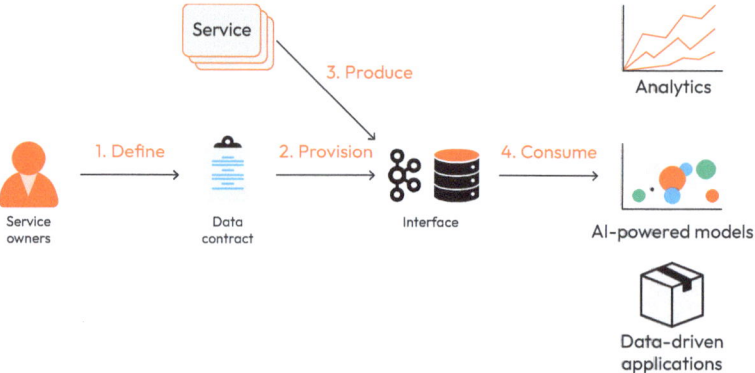

Minimal data contract tooling, built around the interface

It's through the design and applications of these tools that organizations can effectively define responsibilities for data quality, and encourage enhanced collaboration between data producers and consumers. This approach ultimately helps cultivate a data culture wherein high-quality data products are consistently delivered, meeting the requirement standards of data consumers.

Later in the report, you'll explore how to integrate data contracts with the desired cultural shift. But first, let's delve into data mesh as an additional architectural pattern that complements data contracts.

This approach ultimately helps cultivate a data culture wherein high-quality data products are consistently delivered."

Data mesh

Data mesh is a design pattern for building a domain-oriented, decentralized data platform. It focuses not just on the technology but also the social and organizational changes required to achieve this goal. It defines four principles that make up a successful implementation of data mesh:

- **Domain-oriented ownership**: Where data is owned and managed by cross-functional domain teams rather than a centralized data team

- **Data as a product**: Where data is treated as a product that domain teams provide to other teams within the organization

- **Self-serve data infrastructure**: Where tools and platforms are provided to support data producers as they build their data products

- **Federated computational governance**: Where governance is distributed across domain teams and focuses on ensuring that data is used responsibly and ethically

These are all great principles, but implementing some of them may necessitate a substantial organizational and cultural shift. Additionally, data mesh doesn't specify the precise methods for implementing the necessary tooling to support the proposed organizational structures.

Data contracts address this gap by providing an architecture that establishes interfaces where domain ownership is defined and the expectations around the data products are set. This is facilitated through the deployment of a self-service data platform that automates data governance. Therefore, data contracts can be viewed as a precursor to implementing data mesh, or they might suffice as the sole solution for enhancing data quality and maximizing business value derived from data.

Deploying these architectural patterns is imperative for improving data quality, however, it alone isn't sufficient. You also need to change the data culture wherein every individual comprehends the significance of data quality and its role in advancing the organization's objectives.

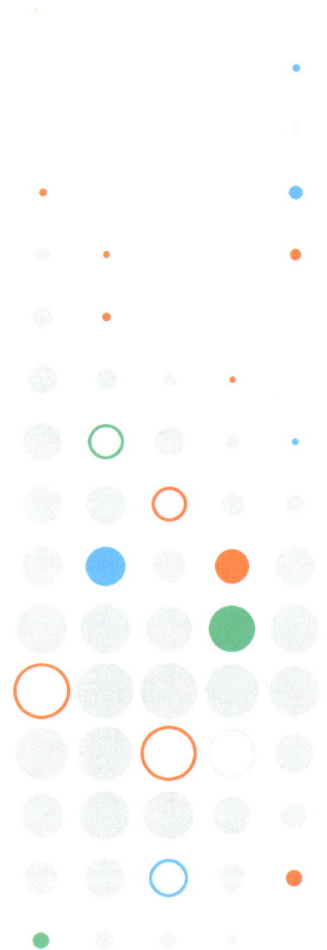

Case studies: Positive impact of data quality

Organizations that prioritize data quality have shown the benefits of that investment, as evidenced by the following case studies.

DoorDash

Today, DoorDash is the largest food delivery platform in the US, but that hasn't always been the case. In January 2018, DoorDash had just 17% share of a super-competitive market, competing against many well-funded competitors.

Food delivery is a low-margin business, so not only did they need to increase market share, but they also needed to increase profitability on every order.

By October 2020, DoorDash had achieved 50% market share, and much of their success has been attributed to their investment in data quality, data platforms, and AI.

DoorDash has invested heavily in a data platform[11] with a focus on:

DoorDash market share

- **Reliability, quality, and SLAs**: DoorDash has recognized the importance of detecting and monitoring the quality of their data and catching any problems as early as possible. This reduces the time to recovery and the associated costs—particularly when dealing with data at scale.

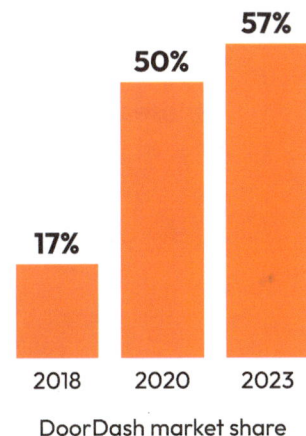

- **Prioritizing trust in data:** DoorDash sees trust in data as essential to the usefulness of data. It shares information early with stakeholders when there has been a breach of SLA, data corruption, or missing data to ensure that they keep their users trust and show that they understand the importance of data quality just as their users do.

DoorDash continues to thrive in their market through the effective use of quality data:

- As of June 2023, their market share had increased to 57%,[12] continuing to grow even as the market itself is decelerating.

- DoorDash's use of data extends to all part of the business, including the heavy use of data science to drive revenue.[13]

- With these foundations in place, DoorDash is looking at some of the ways their business can benefit from the deployment of generative AI.[14]

All of this is possible thanks to their investment in data and data quality.

Checkout.com

Checkout.com is a global payments platform and one of Europe's fastest growing companies. As a company in the financial sector, the quality of their data is critically important for many operational use cases.

As Checkout.com scaled, they found that they had an over-reliance on manual testing and a lack of cross-domain visibility, which resulted in frequent pipeline failures. The data team was struggling to keep up with the alerts, leading to delays in detecting critical incidents with potential business impacts.

To address these challenges, Checkout.com partnered with Monte Carlo[15] to:

- **Increase visibility**: With Monte Carlo, Checkout.com gained visibility into the data health across critical tables, empowering data owners and domain experts. This allowed them to take responsibility for their own data quality, removing the central data team as a bottleneck and reducing the time to recovery from incidents.

- **Adopt incident management**: Once they had increased the visibility and usefulness of their data quality monitors, Checkout.com adopted a comprehensive incident workflow across both data and engineering teams, using tools such as PagerDuty so their engineers could respond quickly and efficiently.

These changes allowed Checkout.com to:

- **Scale beyond manual testing**: Checkout.com now has much greater coverage of their data quality and can identify and resolve issues more quickly through their mature incident management processes.

- **Transform their data culture**: Checkout.com views data as an essential resource that can be used to drive critical operations, fostering a widespread understanding across the organization of the significance of maintaining high-quality data.

This is a great example of looking not just at data quality but also the processes and culture around it, including incident management and decentralized ownership.

Increased visibility into data health

Comprehensive incident workflows

Cultivating a data culture that values quality

Cultivating a data culture that uses quality data to drive business value requires ongoing investment. This might involve implementing initiatives to shift attitudes or recognizing teams that demonstrate desired behaviors.

Let's explore four key areas for integrating data quality into your organization's data culture:

- Adopting a product mindset
- Prioritizing quality over quantity
- Assigning roles and responsibilities
- Embedding data governance

Adopting a product mindset

Encouraging data producers to adopt a product mindset is key to ensuring high-quality data. While data products are often meant for internal use, the principles of product thinking are applicable across various domains, such as developer tooling, which your organization might already be familiar with.

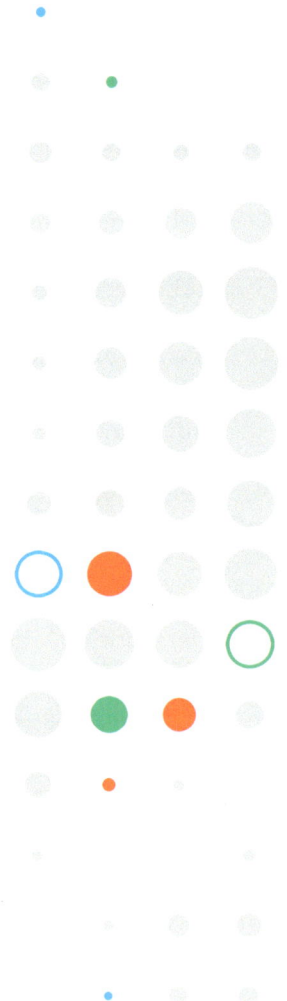

Before creating a data product, it's essential to understand why your organization requires it, who will benefit from it, and what they need and expect from it. These insights stem from data consumers who hold a clear understanding of the business challenges they aim to address through the data product. Articulating these needs to the data producers is vital. Knowing these needs, data producers can appreciate the business value they create by providing this product, giving them a sense of ownership of that value.

These data products should be useful by themselves for driving data-driven applications and services, for example, dashboards and analytics, machine learning models, and internal or external services. They can also be sources for other data products and become part of a supply chain that builds value in each step.

As a principle, these supply chains should be kept as short and efficient as possible, reducing complexity and ensuring that no one team becomes a bottleneck in the use of data. It also keeps the data producers close to the end consumers, so they can see the value of the data they create and feel ownership over the outcomes. For example, if an engineering team can supply the data in the right format to an AI-powered model, they should feel empowered to do so without needing permission or resources from a central data team.

"

Before creating a data product, it's essential to understand why your organization requires it, who will benefit from it, and what they need and expect from it."

To aid with this shift in mindset, you may consider formalizing a *data product manager* role, either as a full-time role or as part of an existing product manager role. They take responsibility for understanding the requirements of the data consumers and work with their product teams to meet those requirements to unlock business value.

The exact implementation of a data product depends on the tools you have available, however, you need to ensure that the expectations are set so users can build on the data product with confidence. This could include the schema, documentation on the purpose and intent of the data, the data quality dimensions that are guaranteed by the owners, and where and how to access the data.

A data catalog could be a good way to collect and display this information, allowing users to find and discover data products that meet their requirements. Some data catalogs also provide data lineage functionality, which allows users to see the supply chain of data products they are building on, and the owners of each part of the data catalog. This increases their confidence in the data and decreases the time to resolution when errors occur. A report from Talend found only 38% of respondents believe their organizations are excellent at tracing back errors into files,[16] and data lineage provides that tracing.

Of course, creating data products takes some effort, and you can't create data products for all your data, but that's okay. This forces you to focus on quality over quantity, which is exactly what you want.

Prioritizing quality over quantity

Since Hadoop came along in 2006 and significantly reduced the cost of storing big data, data engineers have often been focused on how much data they can bring in centrally, with the assumption that they'll use it to create value later. But by prioritizing quantity over quality, many organizations found it took so much effort to use this data that in practice they just couldn't justify it. That left them with *dark data* that was poorly managed and increased the risk of misuse and leaks.

In fact, a report from Seagate found that only 32% of data available to an organization is utilized. That leaves 68% of your data incurring costs, both monetarily and in increased risk, without generating any value.[17]

It's the data producers that are responsible for the quality of their data. But the data consumers are still responsible for supporting that investment. Let's define the roles and responsibilities of both those groups next.

32%

Utilization of available data

Assigning roles and responsibilities

It's only by being clear about roles and responsibilities that diverse groups of people can work together effectively and efficiently to realize the goal of extracting the most business value from data. Let's define the roles of the data generator and the data consumer.

Data consumers

Often, people only think of data consumers as a data practitioner, for example, a data engineer, a BI analyst, or a data scientist. The primary tasks of these professionals require them to consume and work with data, and as such, they are highly reliant on the quality and reliability of that data. But they are not the only data consumers in your organization.

While data consumers cannot be responsible for the quality of the data they use, they do play a major role in shaping that data. They need to be able to articulate their requirements to the data producers and demonstrate the value they can generate through the application of data. They then become accountable for the delivery of that value.

Of course, many data consumers are also data producers. For example, a data engineer that builds a pipeline that transforms and combines data to meet their consumers' requirements, or a service that takes data in, performs an action, and exports the result of that action. So, let's now explore the role of the data producer.

Data producers

According to experts, by 2025, over 463 exabytes (EB) of data will be generated daily worldwide.[18] In many organizations, the roles and responsibilities of data producers have not been well defined. In fact, those producing data often do not even see themselves as data generators, because they spend so little of their time considering how they generate data for other parts of the organization.

If you are going to improve the quality of data at the source, then you need to be explicit about the role of the data producer and their responsibilities. However, there are always trade-offs the data producer needs to make. The consumers might like to have data in a particular format, but that could be expensive for the producer to provide. The consumers might ask for a particular level of performance, but the effort to meet it might be too high compared to the value. Generating this data might also impact the performance of the service they run. So, you need to strike the right balance between the requirements of the data consumers and what the data producers can provide.

This is why the data producer must be the owner of the data contract. Only they can make decisions around these trade-offs, as only they have the full context. They need to be comfortable taking on the responsibility to produce this data and meet the expectations they are committing to. It's the data producers who are going to support and maintain this data over the long term.

"

You need to strike the right balance between the requirements of the data consumers and what the data producers can provide."

A data producer also has responsibility for managing data in line with your organization's policies. This could include categorizing the data, managing access, and ensuring the data is removed when it passes its retention period. They can be supported by tooling that automates this away while also embedding data governance into your organization. Let's look at this in more detail.

Embedding data governance

According to Market Reports World, the global data governance market size valued at USD 2707.97 million in 2022 is expected to expand at a compound annual growth rate (CAGR) of 26.82%[19] during the forecast period, reaching USD 11266.14 million by 2028. This growth highlights the importance of data governance for any organization that aims to maximize the value of its data.

There are many definitions of data governance, and organizations implement it in different ways. Broadly, it is a combination of *people*, *processes*, *standards*, and *technology* that supports and promotes data that is *accessible*, *usable*, *accurate*, *consistent*, *secure*, and *compliant*.

Unfortunately, today, data governance has a bit of a bad reputation. Organizations implement it in a way that it is often seen as a central gatekeeper of data usage. A committee reviews the usage of data, with its main aim being reducing risk, which incentivizes it to slow down or even block the use of data, especially in less-familiar areas.

But data governance is changing. Data that was often governed centrally is now delegated to those teams and domains that produce the data and have the most context on the data. By assigning this responsibility to data producers, there's an increase in the visibility of data governance. This reinforces the importance of data governance.

The role of central data governance is to support those teams and domains through guidance and education, supplied by the experts and available to everyone. This doesn't mean we need to ask our data producers to become experts in data regulation and spend their time implementing our data policies on their data. Instead, we can automate away many of these tasks, and do so rather easily.

One way to do that is to use data contracts, not only for defining the interfaces for your data but also to collect metadata that describes the data. For example, a key part of data governance is the categorization of fields, such as those that contain personal data. By making it easy to categorize that data at the source, it then becomes trivial to build tooling that performs governance tasks on that data, for example, to anonymize or delete that personal data in accordance with your policies or handle access to that data.

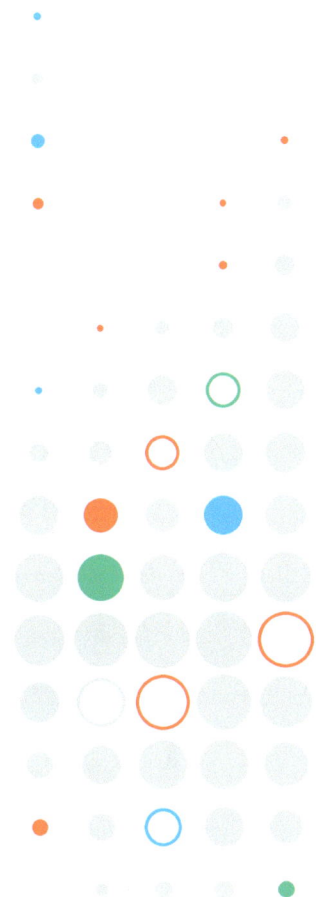

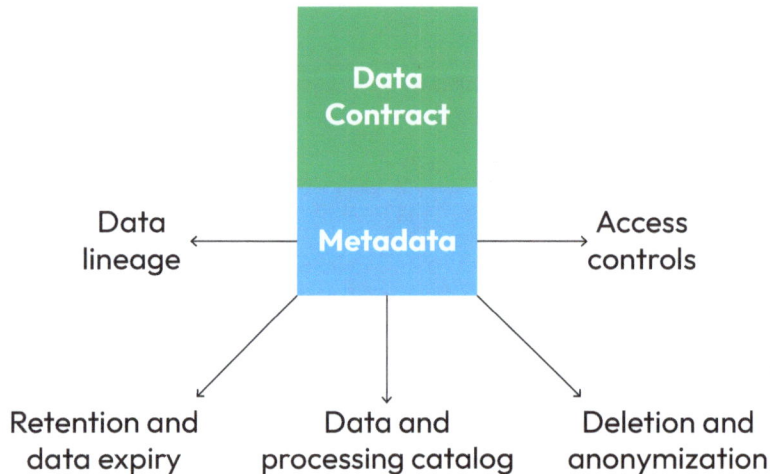

Automating data governance through
the metadata collected in the data contract

By collecting this metadata along with the definition of the data, we're making data producers aware of its importance and ensuring that it stays up to date as the data and its structure evolves.

You can then focus your data governance efforts where it really matters:

- By supporting those who produce and consume data through the deployment of appropriate tooling and automation

- Promoting a data culture that understands the risks and the benefits of using and managing data correctly

- Running programs that promote the data literacy of your organization to encourage better outcomes through the effective use of your data

Conclusion: Embracing a quality-driven data culture

After being introduced to data quality, you saw how critical the quality of data is when building effective AI models, and how you have to make it part of your overall data strategy if your organization is going to succeed in using the latest advancements in AI to drive business value.

But before you can improve your data quality, you need to *measure* it. You can start doing that today by simply asking your data consumers, "*Do you trust your data?*" By doing this regularly through surveys, you'll gain a valuable data point that you can use to measure the quality of the data.

This will inform you about how people *feel* about your data quality, but you can do more. The next step is to gain greater visibility of exactly how big a data quality problem you have, and where you have it. You can do this cheaply by running data quality checks and performing one-off profiling of your data, or you can invest in a data observability platform to continuously monitor each of your data assets and alert you to anomalies in your data.

But if you really want your organization to understand the importance of poor data quality—and the issues caused by it—then you need to start treating your data incidents in the same way that good software engineering teams treat their incidents. Not only does that help increase the visibility of

those issues, but it starts to change your culture to one that doesn't accept these data quality issues must happen, and where data producers and data consumers work together to ensure these issues are prevented at the *source*.

The data source is the only place that can really impact the quality of the data. So, once you have measured and understood the quality of your data, you can start work on providing the support and tools that allow data producers to produce quality data at the source.

There are two emerging and complementary patterns that enable all of this: *data contracts* and *data mesh*. By adopting these as we described in this report, and with support from an effective data platform that embeds data governance, you can make a lasting change to your data culture and become an organization that gets the most business value it can from its unique data.

Once you've done that, there are other areas of data quality that are beyond the scope of this report but are worth considering as the data culture in your organization matures. For example, you can look at improving the quality of the data from your suppliers and other third parties. The principles are the same, but the solutions are different and may involve engaging your procurement team to start prioritizing data quality when choosing new partners. Or, if your organization provides data directly to customers, you can consider how improving the quality of that data can differentiate your offering and increase your revenue.

The resources at the end of this report provide more detail on some of these ideas, but whatever you do next, the foundations you have learned in this report set you up to deliver greater business value through quality data.

About the author

Andrew Jones is a Principal Engineer, author, and a Google Developer Expert. He has over 15 years of experience in the industry, with the first half primarily as a software engineer before moving into the data platform and data engineering space.

He thinks a lot about how to build platforms—particularly data platforms that drive the creation of business value. His work on building performant, efficient, and well-governed data platforms that facilitate the production and consumption of quality data led him to coining and defining *data contracts*, and in 2023 he authored the seminal book on the topic, *Driving Data Quality with Data Contracts*.

Andrew is a regular speaker and writer, and passionate about helping organizations get the most value from data.

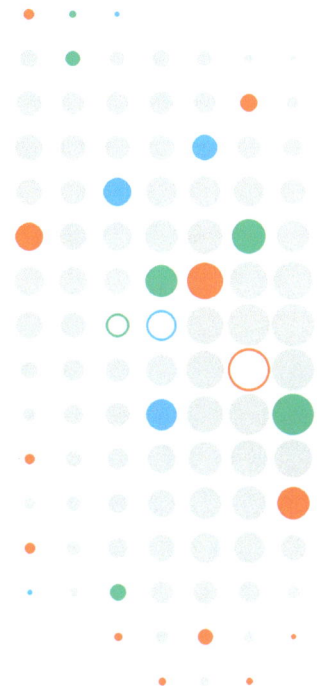

About the technical reviewers

Deepak Bhardwaj

Deepak is a seasoned data mesh and MLOps leader, who is adept at designing cutting-edge data platforms. He champions a decentralized approach, fostering autonomous data teams and empowering organizations to unlock their data assets. Currently leading a data mesh-based platform, Deepak collaborates on structure, governance, and lifecycles for enhanced team autonomy. Specializing in MLOps, he optimizes platforms for the entire machine learning lifecycle, emphasizing cloud-native technologies.

Piotr Czarnas

Piotr, recognized as one of the "Top Data Quality voices on LinkedIn," serves as the CEO of DQOps, a leading data quality monitoring company. With over 20 years of expertise in the field, he is a seasoned professional dedicated to addressing data quality challenges. Piotr's leadership has been instrumental in steering his company towards excellence, providing robust solutions for monitoring and enhancing data quality. His extensive experience positions him as a thought leader in the industry, contributing valuable insights and strategies to the broader community.

Amit Sharma

Amit is currently serving as a Data Product Manager at Wefox, specializing in addressing challenges related to data observability, data quality, and data governance. Through the implementation of effective processes, the data reliability index under his management has increased by 10%, reaching a commendable 4.6/5 from 4.2. With a focus on enhancing reliability and trustworthiness, Amit aims to accelerate the adoption of data products within organizations, thereby boosting ROI for the data team. His background includes roles in testing applications, both front end and back end, and developing automation test frameworks. Proficient in various technologies, such as Python, Java, Selenium, Cypress, Pytest, Test NG, DBT, and pandas, Amit ensures data quality through robust frameworks and methodologies.

John Thomas

John Thomas, a data analytics architect and dedicated book reviewer, combines his passion for data and technology in his work. He has successfully designed and implemented data warehouses, lakes, and meshes for organizations worldwide. With expertise in data integration, ETL processes, governance, and streaming, John's eloquent book reviews resonate with both tech enthusiasts and book lovers. His reviews offer insights into the evolving technological landscape shaping the publishing industry.

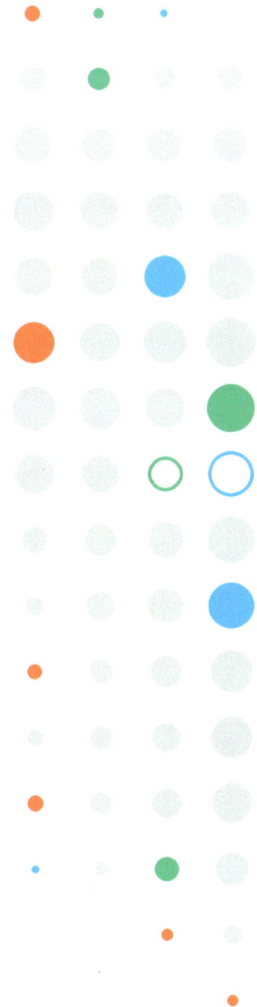

Additional reading

You can learn more about data quality issues from the following sources:

- Askham, Nicola. "Navigating Data Mesh and Evolving Data Governance: A Practical Guide." *NikolaAksham.com*, November 30, 2023. https://www.nicolaaskham.com/blog/2023/11/30/navigating-data-mesh-and-evolving-data-governance-a-practical-guide. This guide discusses how data governance is evolving alongside emerging architecture patterns such as data mesh.

- Colsey, Jack. "'The dashboard looks broken!': How should data teams respond to incidents?" *Incident.io*, November 21, 2023. https://incident.io/blog/incident-management-for-data-teams. It discusses how data teams can start following a mature incident response process.

- Dehghani, Zhamak. *Data Mesh: Delivering Data-Driven Value at Scale*. O'Reilly Media, 2022. The seminal book on data mesh by the person who coined it, discusses each of the principles in detail, including data products and federated data governance.

- Gavish, Lior, and Moses, Barr. "Data Observability: Reliability in the AI Era." *Monte Carlo*, November 27, 2023. https://www.montecarlodata.com/blog-data-observability-reliability-in-ai-era/. This book discusses the importance of mature data engineering if you want to meet your AI objectives.

- Hawker, Robert. *Practical Data Quality: Learn practical, real-world strategies to transform the quality of data in your organization*. O'Reilly Media, 2023. This is a great practical guide to improving the quality of data at any organization. It includes complete definitions of data quality metrics with examples.

- Hellerstein, Joseph M., et al. *Principles of Data Wrangling: Practical Techniques for Data Preparation*. O'Reilly Media, 2017. This includes good advice on how to analyze data that's available and the quality of it, including using techniques such as profiling.

- Hu, Kevin. "The importance of data quality of product-led companies." On medium. com, August 17, 2022. https://kevinzenghu.medium.com/the-importance-of-data-quality-for-product-led-companies-661d7d50d3b9. It highlights the many different areas of a business that are directly affected by poor-quality data.

- Jones, Andrew. *Driving Data Quality with Data Contracts: A comprehensive guide to building reliable, trusted, and effective data platforms.* Packt Publishing, 2023. This book discusses in detail how to use data contracts to change your data culture to one that produces and consumes quality data, supported by an effective data platform.

- Klein, Zack. "Data Contracts in the Modern Data Stack." *Whatnot Engineering* on medium.com, July 13, 2023. https://medium.com/snowflake/data-contracts-in-the-modern-data-stack-d42cb2442dbd. This resource shows how Whatnot is using data contracts at their organization to produce more consistent, better-quality data at scale.

- Madsen, Laura B. *Disrupting Data Governance: A Call to Action.* Technics Publications, 2019. This book introduces a more modern take on data governance, with a focus on supporting people across the organization.

- Moses, Barr, Lior Gavish, and Molly Vorwerck. *Data Quality Fundamentals.* O'Reilly Media, 2022. This guide showcases the problems caused by poor-quality data and offers practical solutions.

- Petrella, Andy. *Fundamentals of Data Observability: Implement Trustworthy End-to-End Data Solutions.* O'Reilly Media, 2023. This practical guide details how to gain greater visibility of data and its usage.

Bibliography

1   Informatica, January 31, 2024. "New Research from Informatica Reveals More Data Leaders Plan to Capitalize on Generative AI, but Data Quality Remains the Key Obstacle to Adoption." https://www.informatica.com/about-us/news/news-releases/2024/01/20240131-new-research-from-informatica-reveals-more-data-leaders-plan-to-capitalize-on-generative-ai-but-data-quality-remains-the-key-obstacle-to-adoption.html.

2   Datiku 2023. "AI, Today: Insights From 400 Senior AI Professionals on Generative AI, ROI, Tech Stack, & More." https://content.dataiku.com/dataiku-databricks-survey.

3   Twilio Segment, 2023. "The State of Personalization 2023." https://segment.com/pdfs/State-of-Personalization-Report-Twilio-Segment-2023.pdf.

4   Anaconda, 2022. "2022 State of Data Science." https://www.anaconda.com/resources/whitepapers/state-of-data-science-report-2022.

5   Jamie Galvin, June 21, 2022. "Unity Software: It's Not All Doom And Gloom." https://seekingalpha.com/article/4519609-unity-software-stock-not-all-doom-and-gloom.

6   Magnus Eriksen, January 04, 2023. "Data Quality: How Bad Data Can Impact Your Business." https://www.quintly.com/blog/data-quality-how-bad-data-can-impact-your-business.

7   Monte Carlo, 2023. "The State of Data Quality: Data Leader Strategies & Benchmarks." https://resources.montecarlodata.com/ebooks/data-quality-survey-1.

8   Solomon Radley, 2021. "Data Integrity Trends: Chief Data Officer Perspectives in 2021." https://www.precisely.com/app/uploads/2021/06/Data-Integrity-Trends-2021-Corinium-Intelligence.pdf.

9   Anne-Laure Thieullent, Valérie PERHIRIN, et.al. 2020. "The data-powered enterprise." https://www.capgemini.com/wp-content/uploads/2021/02/Data-powered-enterprise_Digital_Report-4.pdf.

10   Andrew Abraham, February 25, 2021. "Highlights from Our 2021 Global Data Management Research." Experian. https://www.experian.com/blogs/news/2021/02/25/highlights-2021-global-data-management-research/.

11   Sudhir Tonse, September 2025, "How DoorDash is Scaling its Data Platform to Delight Customers and Meet our Growing Demand." https://doordash.engineering/2020/09/25/how-doordash-is-scaling-its-data-platform/.

12   Earnest Insights, June 13, 2023. "DoorDash and UberEats pushing GrubHub out of Restaurant Delivery." https://www.earnestanalytics.com/insights/all-posts/doordash-ubereats-pushing-grubhub-out-of-restaurant-delivery.

13   CK Gustafson, November 28, 2023. "How DoorDash uses Data Science to Drive Revenue." Mostly Metrics. https://www.mostlymetrics.com/p/how-doordash-uses-data-science-to.

14   Alok Gupta, April 26, 2023. "DoorDash identifies Five big areas for using Generative AI." https://doordash.engineering/2023/04/26/doordash-identifies-five-big-areas-for-using-generative-ai/.

15   Tim Osborn, January 31, 2023. "How Checkout.com Achieves Data Reliability at Scale with Monte Carlo". Monte Carlo. https://www.montecarlodata.com/blog-how-checkout-com-achieves-data-reliability-at-scale-with-monte-carlo/.

16 Talend. April, 2019. "The 10 habits of highly effective organizations." https://www.talend.com/resources/data-trust-readiness/.

17 Seagate. 2020. "Rethink Data." https://www.seagate.com/in/en/our-story/rethink-data/.

18 Edge Delta, March 11, 2024. "Breaking Down The Numbers: How Much Data Does The World Create Daily in 2024?" https://edgedelta.com/company/blog/how-much-data-is-created-per-day.

19 MarketReportsWorld, September 25, 2023. "Data Governance Market Insights [2023-2030] | The Ultimate Industry Handbook." https://www.linkedin.com/pulse/data-governance-market-insights-2023-2030-ultimate/.

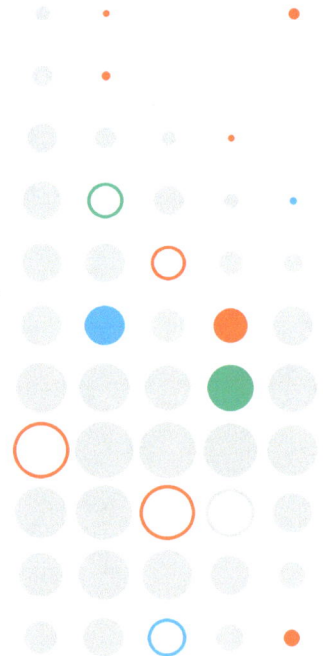

‹packt›

www.packtpub.com

Subscribe to our online digital library for full access to over 7,000 books and videos, as well as industry leading tools to help you plan your personal development and advance your career. For more information, please visit our website.

Why subscribe?

- Spend less time learning and more time coding with practical eBooks and Videos from over 4,000 industry professionals

- Improve your learning with Skill Plans built especially for you

- Get a free eBook or video every month

- Fully searchable for easy access to vital information

- Copy and paste, print, and bookmark content

Did you know that Packt offers eBook versions of every book published, with PDF and ePub files available? You can upgrade to the eBook version at packtpub.com and as a print book customer, you are entitled to a discount on the eBook copy. Get in touch with us at customercare@packtpub.com for more details.

At www.packtpub.com, you can also read a collection of free technical articles, sign up for a range of free newsletters, and receive exclusive discounts and offers on Packt books and eBooks.

Other Books You May Enjoy

If you enjoyed this book, you may be interested in these other books by Packt:

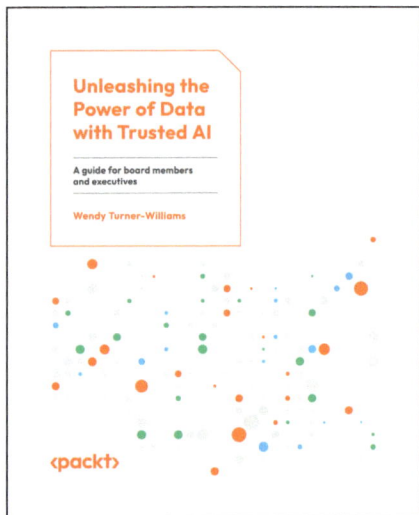

Unleashing the Power of Data with Trusted AI

Wendy Turner-Williams

ISBN: 978-1-83546-789-3

- Navigate ethical considerations and comply with data regulations effectively
- Elevate data quality and enhance data literacy within your organization
- Craft effective AI strategies for data analytics processes
- Explore real-world case studies showcasing the tangible benefits of trusted AI
- Optimize decision-making processes by harnessing AI-driven insights

Packt is searching for authors like you

If you're interested in becoming an author for Packt, please visit authors.packtpub.com and apply today. We have worked with thousands of developers and tech professionals, just like you, to help them share their insight with the global tech community. You can make a general application, apply for a specific hot topic that we are recruiting an author for, or submit your own idea.

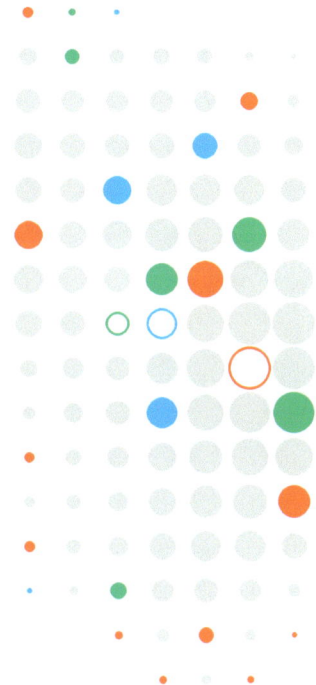

www.ingramcontent.com/pod-product-compliance
Lightning Source LLC
Chambersburg PA
CBHW041933220326
41598CB00058BA/838